Nature's Children

COWS

by Frank Puccio

Grolier Educational

FACTS IN BRIEF

Classification of cattle.

Class:	*Mammalia* (mammals)
Order:	*Artiodactyla* (cloven-hoofed mammals)
Family:	*Bovidae* (antelope, cattle, sheep, and goats)
Genus and species:	*Bos taurus* (European breeds) and
	Bos indicus (Indian breeds)

World Distribution. Worldwide.

Habitat. Domestic cattle are usually raised on open range or on pasture land for grazing, and they often have a barn or other shelter.

Distinctive physical characteristics. Some have horns, distinctive split (two-toed) hooves.

Habits. Cattle are ruminants, or cud-chewers, that graze on pasture and live in herds.

Diet. Grass and other plants that grow on the pastures on which they graze; hay; grain; commercial feed.

Library of Congress Cataloging-in-Publication Data

Puccio, Frank, 1958-
    Cows / Frank Puccio.
        p.   cm. — (Nature's children)
    Includes index.
    Summary: Gives information on the physical characteristics,
behavior, natural habitats and uses of cows, steers, and oxen,
focusing on the types that might be seen on farms, zoos, and at
rodeos.
    ISBN 0-7172-9117-0 (hardbound)
    1. Cattle—Juvenile literature.   [1. Cattle.]   I. Title.
II. Series.
SF197.5.P83   1997
626.2—dc21

97-5973
CIP
AC

This library reinforced edition was published in 1997 exclusively by:

 Grolier Educational
Sherman Turnpike, Danbury, Connecticut 06816

Set ISBN 0-7172-7661-9
Cows ISBN 0-7172-9117-0

Contents

The gentle, lovable cow is one of the most useful of all domestic animals.

What comes to mind when you hear the word cow? Do you think of a large animal that spends its day eating grass and giving off an occasional moo? Or do you think of ice cream, cheese, or any of the other products that come from cows?

To most of us cows merely supply some of our most important foods. But there is a lot more to cows than this.

In many parts of the world, for example, cows—or their near relatives—often take the place of farm equipment, just as they have for thousands of years. They plow fields, carry people to the harvest, and even cart goods to market.

Cows do other jobs as well. They have been widely used to clear land, especially for back-breaking tasks like pulling tree stumps out of the ground. Cows have even been used for transportation. A look at pictures of pioneers on the famous Oregon Trail will show you teams of oxen pulling people and their belongings to the West.

Cow, by the way, is just one of the many names referring to this very useful animal. When correctly used, for example, cow means only the female milk giver. The word cattle, on the other hand, includes cows, calves (very young cattle), bulls (males), steers (neutered males), heifers (young females), and oxen.

Cows eat . . . and eat . . . and eat. Everything is chewed and then rechewed many times over before it is finally digested.

Did You Know That . . .

Cattle are among the most important of all farm animals, which explains why people have been raising them for almost 10,000 years.

Just why are cattle so important? To begin with, much of what we eat comes from cattle, including milk and cheese, yogurt and ice cream. Besides these there are steak, hamburger, and many other meat products. But that is not all. We also get leather, soap, glue, and even medicine from cows.

Another interesting fact is the way cattle eat. In order to digest their food properly, these grass eaters chew their food twice. After chewing it the first time, the cow swallows the food. After a while the animal brings the food up from its stomach and chews it again. This cud, as it is called, is what cows seem to be chewing all the time.

Here's another fact. Depending on what they are used for, cows are put into one of three groups. Beef cattle, the first group, are raised for their meat. Dairy cattle are raised for their milk. The third group is made up of dual-purpose cattle, which can be used for both meat and milk.

What They Look Like

Cows have large, heavy bodies. Their long tails are used for keeping flies away. Cow feet are protected by curved coverings of horn, called hoofs, which are split, or cloven.

Fully grown cattle are very muscular. Adult dairy cattle usually reach about 5 feet (1.5 meters) tall and weigh from 900 to 2,000 pounds (410 to 900 kilograms). Bulls often weigh even more.

Cattle hides are usually black, white, light brown, or a combination of these colors. Although we generally think of cattle hair as fairly short, it usually grows longer in the winter in order to keep the animals warm. One breed, the Galloway, actually has long, shaggy hair all year. This heavy coat protects the animal from the cold weather in its native Scotland.

All of the females have special organs called udders, which hold milk. The udders hang down between and just in front of the cow's hind legs. Dairy cows are bred to have particularly large udders in order to make milk production more efficient.

Cattle have large bodies and long tails.

9

Beef Cattle

The meat that comes from cattle is called beef. Pot roast, steak, and hamburger are types of beef.

Beef cattle usually are raised on land that is not rich enough to produce food crops such as wheat or vegetables. Cattle are not fussy eaters and, in fact, usually do quite well on whatever plants happen to be present on the land. However, ranchers (people who raise cattle) generally have to provide food for them during the cold winter months when the grasses die off or become buried under snow.

Over the years beef cattle have been carefully bred to grow and mature to adulthood quickly. This too saves ranchers money. The faster an animal reaches the desired size and weight, the faster it can be sold. And less time spent raising an animal means less money spent on cattle feed and other expenses.

People eat not only the steaks, roasts, and hamburgers that come from cattle, but other parts as well. In fact, just about every part of the animal—including the internal organs such as the liver, kidneys, and stomach—is used. And what is not used for human food is made into pet food or some other product.

There are more beef cattle in the United States than any other type of cattle.

11

Breeds of Beef Cattle

North America, which produces much of the world's beef, raises five major breeds of cattle. It raises them in such great numbers that in the United States, beef cattle outnumber dairy cattle by about four to one.

The Angus, one of the most famous of all beef cattle breeds, is prized for its excellent meat. Originally developed in Scotland, it was brought to the United States in 1873.

The Brahman was bred in this country from cattle imported from India. It tolerates hot weather and is able to resist diseases better than many other breeds.

Charolais cattle, originally from France, are particularly large. Ranchers often crossbreed Charolais with other breeds in order to produce meatier animals.

The Hereford is one of the most common breeds and is raised everywhere from New England to the Far West. It can thrive in a wide range of temperatures and is especially good at enduring the bitter cold winters of places like Montana and western Canada.

The shorthorn was developed in England during the late 1700s. It has become particularly popular in recent years—as feed prices go up and up—because its calves mature quickly.

The Angus is one of the most famous of all breeds raised in North America.

The Beginnings of the Cattle Industry

According to pictures engraved on their tombs, the ancient Egyptians raised cows and put them to work pulling plows. But cattle raising on a really large scale didn't occur until the 1800s in North America.

Cattle are not native to North America. It is likely that they came here with the Vikings during their exploration of the continent about 1000 AD. But the first large group of cattle arrived with Christopher Columbus during his second voyage to America in 1493. These animals were intended for use in the Spanish colonies in America. Not too long after that, cattle could be found in many Spanish communities throughout the Caribbean.

Several years later, when Hernando Cortés conquered Mexico for Spain, he brought descendants of these cattle to the mainland. From there cattle spread northward into Texas and the Southwest.

Other European colonists brought cattle to North America as well. Soon cattle could be found just about everywhere, from Canada to Mexico and from New England to the Pacific.

The Texas longhorn developed from the cows brought to Mexico by the conqueror Hernando Cortés.

The American Cowhand

Unlike so many other folk heroes, cowhands were real people who played an important role in building the American West.

In the 1800s, from Mexico to Canada, cattle were raised on the open range. There the animals were let loose onto unfenced land to graze and fend for themselves. This system saved money for the rancher. But it also created hard work —and a permanent place in American legend—for the cowhand.

Life for these rugged men was very tough. Most of their time was spent outdoors, regardless of weather or season, health or holiday. The hours were long— sometimes 24 hours a day—and the work itself was difficult and demanding.

Cattle had to be rounded up, counted, and branded with marks to show ownership. Cows had to be assisted during birthing, and calves had to be kept from roaming off. There were predators to deal with, as well as terrible storms and blizzards. There also was the occasional cattle thief—or rustler—to chase off. To do all this, cowhands had to know how to ride, rope, take care of animals, and even handle a gun. The reality of the job was a far cry from the glamorous life dreamed about by countless youngsters.

The American cowhand is part reality and part folklore.

The Great Cattle Drives

Each year during the 1860s through the 1880s ranchers rounded up their cattle and moved them to places far away where they could load them onto the railroad. From there the cattle were shipped to the stockyards and slaughterhouses of Chicago.

The most famous of these long cattle drives took place along the Chisholm Trail. The trail was blazed in 1866 when Jesse Chisholm drove a wagon through what is now Oklahoma to his trading post near Wichita, Kansas. Cowhands soon followed, using the ruts made by Chisholm's wagons as guides. In time herds of 4,000 cattle were following a trail that ran all the way from the Mexico-Texas border to Kansas. Soon other trails were in use as millions of cattle made their way from the range to the market.

The long, dangerous drives often lasted two to three months and covered as much as 1,000 miles (1,600 kilometers). Along the way cowhands faced everything from rustlers to mountain lions to floods.

By the late 1890s the railroad system had spread out, coming closer to the ranches it served. Soon the great cattle drives were a thing of the past.

Driving cattle today is nothing compared with the great cattle drives of the 1860s and 1870s.

Branding Cattle

Back in the days of the open range cattle from different ranches often grazed together for months at a time. Without brands (the marks that showed exactly who owned a particular animal) ranchers would never have been able to tell which livestock belonged to whom.

In the Old West—and even today—calves are branded when they are just a month or two old. The animals first are gathered together. (This is one of the famous "roundups" of Western legends.) Then the calves are roped, held down, and marked with a hot iron. One end of the iron is twisted and bent to form the shape of the ranch's brand. When the iron touches the calf's hair, it burns a scar, leaving the ranch's brand on the animal forever.

The first brands were fairly simple lines made with a straight rod. As time passed, however, brands soon became a kind of language that cowhands and cattle people immediately understood. A letter T with a short straight line over it, for example, would stand for the "T-Bar Ranch." A slanted T would stand for the "Running T."

Cowhands take care of calves as well as rope and brand them.

Rodeos

Rodeos began as a way for cowhands to relax out on the open range or on a long cattle drive. At the end of the day (or maybe on a day off) cowhands would gather and test themselves at skills that were central to a cowhand's way of life. Riding horses, roping calves, and wrestling steers to the ground were all part of the game. Soon cowhands were holding informal competitions to see which of them was best at each skill.

With time towns and counties got into the act. Prizes were given, and the entire atmosphere resembled a carnival or state fair. Eventually there were even professional competitors traveling from rodeo to rodeo. These cowhands competed for money and for the silver or gold belt buckles that became trophies of a rodeo championship.

Today rodeos have five major events, all of which date back to the skills of the old-time cowhand: bare-back bronco riding, saddle bronco riding, calf-roping, steer wrestling (or bull dogging), and bull riding. Through events such as these the ways of the old cowhands remain alive for all of us to enjoy.

Bull riding is one of the original rodeo events, and it still is one of the most popular.

Cattle Ranching Today

On most ranches cattle raising follows the same predictable pattern. Calves are born in the spring of the year. At the age of a month or so the calves are branded and set out to graze. The open range, of course, is long gone so today's cattle graze on land that is owned or rented by the ranchers themselves. There, on large areas of fenced-in grassland, the cattle eat and grow during the spring and summer.

The cattle eat and rest, protected from mountain lions, coyotes, and other predators by herd dogs or the ranchers themselves. The ranchers check on the cattle often and stick to a busy schedule of chores. Mending fences, planting hay, and tending sick or injured cattle are all part of the day's work.

By autumn the calves—called yearlings now—have reached 450 to 650 pounds (200 to 300 kilograms). On some ranches the cattle will remain for another year or so. But most yearlings are sold to other ranchers who fatten the cattle with extra protein, vitamins, and food supplements. After three to five months the cattle reach 1,100 pounds (500 kilograms) and are ready for the meat-packing plant.

Today cattle are fattened up with special feeds.

Dairy cows provide some of our most important foods.

Dairy Cattle

In the days of ancient Egypt and Rome the same breeds of cattle did everything. They plowed fields and hauled wagons, while also providing milk and meat.

In time farmers began breeding their cattle to produce certain qualities. They would mate the best milk-producers, the best meat-producers, and so on. In this way specialized breeds were developed.

Today's dairy cattle are the result of hundreds of years of careful breeding. Because of this the average dairy cow produces 80 glasses of milk each day. That adds up to almost 15,000 pounds (7,000 kilograms) of milk each year.

Modern dairy farms are far more efficient than the farms of the past. Machines, for example, now do most of the milking because machines are faster and waste less milk than people do. They also are cleaner, safer, and gentler than human hands.

Beyond this today's dairy cows get a carefully planned diet of hay, grasses, and grains. They are also given vitamins and supplements to keep milk production high. Some farmers actually use computers to figure out what and how much to feed their cows.

The Great Milk Factory

Most mammals produce just enough milk for their own babies. Dairy cows, however, are different. During its lifetime the average calf needs only about 66 gallons (250 liters) of its mother's milk. Since the mother produces 1,585 gallons (6,000 liters) during the ten months after the calf's birth, there's a lot of extra milk to put to good use elsewhere.

In order to produce so much milk, a cow eats a great deal of food each day—9 pounds (4 kilograms) of hay and 35 pounds (16 kilograms) of silage, which is a mixture of grasses and grains. It also manages to find room for 23 pounds (10 kilograms) of mixed grains, salt, vitamins, and minerals. And it washes all of this down with 15 gallons (57 liters) of water.

Besides eating so much, cows know how to get the most out of their diet. As you know, a cow chews her food twice, the first time as raw food and the second as round lumps called cud. After being chewed again and again, this cud finally goes back to the cow's stomach for complete digestion. The point of all this chewing and digestion is to give the cow all the nutrients it needs. But it also helps the cow make milk—lots of it.

Today's dairy cow is an amazing milk factory that is able to produce up to 80 glasses of milk each day.

Dairy Cattle Breeds

Of the dozens of dairy cattle breeds in the world, a few have proven themselves to be the clear favorites. In North America five major breeds of dairy cattle represent the majority of the milk industry.

Holsteins are by far the most common of all dairy cattle in the United States and Canada. Their popularity is based on the ability to produce far more milk than any other kind of cow. Large animals with black-and-white spotted markings, they originally came to this continent from Holland.

The other four breeds account for most of the remaining dairy cattle in North America. The Brown Swiss is one of the oldest breeds. Raised in other countries for its meat, in North America it is raised primarily for its milk.

The Ayrshire is a rugged breed that adapts well to hilly places. In milk production it ranks third after the Brown Swiss and the Guernsey.

Guernsey and Jersey cows originally came from small islands in the English Channel. Guernsies are famous for their creamy, yellow milk. Jerseys are actually a bit smaller than most other dairy cattle, but their milk is rich in butterfat.

Modern machinery now does the milking that was once done by hand.

Dairy Farming Yesterday and Today

European colonists who came to America during the 1600s and 1700s brought dairy cattle with them. The cows bred successfully and quickly became part of most family farms.

Later, as Americans moved west, so did their cattle. During the 1700s and 1800s farm families on the western frontier depended on their dairy cattle for milk, butter, and cheese. These same cows also pulled plows and hauled wagons and, when the time came, became meat for the table.

The growth of cities changed all this. Suddenly there were huge populations of people who wanted milk and milk products but who were unable to raise their own cows. So dairy farms sprang up nearby, where farmers would raise cows and sell the milk to companies which, in turn, would sell it in the city.

Today farmers still raise cows and sell the milk. But the process is faster and more efficient. For example, most dairy farmers now belong to organizations called cooperatives, which help them get the best possible prices for their goods. In addition farmers use modern techniques and equipment that help them get the most out of each cow.

Modern Milking

In the old days cows were milked by hand. This took a long time and was no easy task. Today, however, machines do the job, usually twice each day. Not only can the machines milk several cows at once, but the job is done in a faster, easier, and far cleaner way!

A typical milking machine has four cups, one for each of the four nipples on a cow's udder. As the farmer gets ready to start milking, the cow's udder is carefully washed with disinfectant. This keeps the process clean and eliminates germs and disease.

Then one cup is hooked to each nipple. When the machine is turned on, a gentle suction draws the milk out of the cow. In less than five minutes it is all over, with not the slightest pain or discomfort to the cow. This goes on twice each day for 10 months each year. At that point the cow will stop giving milk unless it becomes pregnant and gives birth to another calf.

After milking, the milk passes through steel or glass pipes to a refrigerated tank. Here the milk is kept fresh and prevented from spoiling until it is picked up by a truck from the farmer's cooperative. Meanwhile, the milking machine is carefully cleaned and made ready for the next session.

Every day or two raw milk is trucked from the farm to the processing plant.

Processing the Milk

Every day or so the milk is trucked from the farm to the dairy. Here it is tested, processed, and set up for delivery to stores.

The first step in processing milk is pasteurization. Like anything natural, raw milk—the milk straight from the cow—may contain bacteria, which could cause illness. So all milk that is to be sold is treated by a process first developed by the French scientist Louis Pasteur back in the 1800s. The milk is heated to at least 160° Fahrenheit (72° Celsius) for about 16 seconds and then cooled. Pasteurizing kills the dangerous bacteria and makes the milk safe to drink.

At this point the milk could be sold, but it usually goes through a second process, homogenization. This makes milk smoother by breaking down the natural milkfat into smaller particles.

Once this is done some of the milkfat can be removed from the milk to make 2% or 1% milk, skim milk, or even nonfat milk. At the same time nutrients like vitamin D or calcium may be added. Finally, when all this is done, the milk is packaged and sent to stores.

Other Products from Milk

Not all raw milk ends up in milk cartons. In fact, less than 40% is sold as milk. The rest is used in making milk-related products.

To make butter, for example, cream (the part of the milk richest in milkfat) is pasteurized and then churned. Stirred again and again, the milkfat becomes more and more solid until it turns into rich, creamy butter.

Producing cheese is somewhat more complicated. Cheese is made by adding bacteria to milk, which then causes soft curds to form. The liquid milk is then removed from the curds, either by squeezing them by hand or by machine. It is the curds that later become cheese.

Besides butter and cheese, milk products include buttermilk, evaporated milk (which is canned, so it does not need to be kept cold), cottage cheese, yogurt, and ice cream. Each of these is made in a slightly different way.

Dairy cows like this brown Swiss produce dozens of products in addition to milk.

Ice Cream

Try this one on for size. On July 24, 1988, Palm Dairies of Alberta, Canada, created the largest ice cream sundae ever made. With 44,687 pounds (20,270 kilograms) of ice cream, over 8,818 pounds (4,000 kilograms) of syrup, and a whopping 537 pounds (243 kilograms) of toppings, it weighed in at just under 55,115 pounds (25,000 kilograms).

All of that ice cream was made the same way that all ice cream is made. Milk and sugar are blended together and then pasteurized for safety. Next, the mixture is homogenized, stored, and cooled for a while. At this point flavorings go in—vanilla, chocolate fudge, liquid fruit, whatever is wanted.

Then it's off to the freezer, where the ice cream mix is chilled. At the same time, air bubbles are added to make the mixture smooth and creamy. After this the chunky flavorings go in—nuts, pieces of fruit or cookies, chocolate chips, or even cookie dough. Then, before it can melt, the ice cream gets packed into containers.

Ice cream is probably the world's favorite milk product.

Dual-Purpose Cattle

Although some farmers and ranchers prefer specialized animals, not all cattle are only beef cattle or just dairy cattle. In fact, a number of breeds have been created just so they can be both meat producers and milk producers.

Milking shorthorns, for example, produce large amounts of both beef and milk. Although they are found all over North America, these cattle are especially popular among farmers in the eastern United States and in the Midwest. Milking shorthorn calves grow especially fast. This makes them inexpensive to raise, especially since they can be shipped to meat-packing plants at an earlier age than other cattle.

Another popular dual-purpose breed is the red poll. (A polled animal is one without horns.) This breed was developed in England by crossing a horned breed with a hornless one. Not as large as the milking shorthorn, it is used as a dairy breed more often than as a meat-producing one.

Oxen

Oxen have a special place in the history of domestic animals, with a reputation for strength, hardiness, and spirit. Physically they are bigger than cattle, with large heads, heavy bodies, and powerful muscles. These muscles have helped oxen do everything from pull plows to haul wagons.

Although few oxen are used in North America as work animals any more, many people continue to think fondly of the huge, powerful creatures. A county or state fair usually isn't complete without an ox-pulling contest or two. In these, oxen (usually two to a team) strain to see which of them can move the greatest weight. Trained to work and pull together, the animals tug at sleds loaded with anywhere from 500 to 3,000 pounds (225 to 1,350 kilograms) of metal or concrete weights.

When they are not involved in occasional pulling contests, oxen live a pretty good life. They are well-loved, well-fed pets that can usually be found grazing in a big, green meadow.

India's Sacred Cows

No country in the world has more cows than India. Yet very few of these animals are raised for their beef. This may seem strange since so many people in India are underfed. But to those who know India, this is no mystery. In India, about 80 percent of the people practice the Hindu religion, which believes that animals have souls just as people do. Some Hindus even treat certain animals as sacred beings deserving of special care. One such animal is the cow. For this reason cows are rarely raised for slaughter in India, and the thought of eating meat from a cow is an unpleasant—or even shocking—one to many people there.

Although they are not raised for meat, cattle are still very important to Indian farming. Many of the farmers are poor, unable to afford the expensive, modern machinery available in other parts of the world. With tractors and other equipment difficult—or even impossible—to buy, Indian farmers often turn to cattle to do their heavy labor. So the sight of a cow or bull plowing a field is not at all unusual in India.

Oxen like this Devon ox do many jobs.

In India, where cows are sacred, cattle work and provide milk but are never slaughtered for meat.

Ranching around the World

Although North America is the world's most famous cattle-producing country, other nations and areas raise beef and dairy cattle, too.

Some of the world's largest beef cattle ranches, for example, are found in South America. Argentina, South America's second-largest country, has long been famous for its ranches and cattle industry. Found mostly on the grassy plains of Argentina's pampas, the country's ranches are run by rugged cowhands called gauchos. Over the years gauchos have become part of Argentine folklore in the same way that cowhands have become part of American folklore.

Venezuela is another South American ranching country. Its cattle ranches are located in a flat region called the llanos, and Venezuelan cowhands are called llaneros.

The Australian cattle industry rivals the one in North America in size and development. For years Australian ranchers have carved lives for themselves on land as harsh as anything found in the American Southwest.

Bullfighting

In almost any city in Spain Sunday is the day for bullfighting, a colorful tradition that goes back hundreds of years. These contests pit a human bullfighter—the matador—and several assistants against a 1,000-pound (450-kilogram) bull that has been carefully bred to display fierceness, strength, and courage.

Everything about the bullfight—even today—is done according to carefully maintained traditions. The matador's colorful outfit—the suit of lights, as it is called—is the same as it was centuries ago. The same is true for the pattern of the bullfight, which has carefully arranged stages and movements. One of these comes when the matador uses a cape to encourage the bull to charge. Sweeping the cape in graceful motions, the matador brings the bull closer and closer in ever more dangerous moves. In the end, when the matador finally plunges a sword through the bull's neck and into its heart, the crowd cheers the courage of both the matador and the defeated bull.

Words to Know

Brand   The mark on an animal that identifies its owner.

Branding iron   The tool used to brand cattle.

Bulls   Adult male cattle.

Calves   Young cattle.

Cattle drive   The movement of cattle to places where they could be put on the railroad and shipped to the stockyards or slaughterhouses.

Cows   Adult female cattle.

Cud   A lumpy mass of partially chewed food that is rechewed by cattle.

Heifers   Young female cows.

Homogenizing   To process milk by breaking up and distributing the milkfat more evenly.

Matador   A bullfighter.

Oxen   Large cattle used as work animals.

Pasteurizing   To process milk by heating it in order to kill possibly harmful bacteria.

Roundup   The gathering of cattle for branding, movement to new pastures, or travel.

Silage   Feed for cattle.

Udder   The organ on a cow that holds her milk.

INDEX

Cover Photo: Lynn M. Stone
Photo Credits: Norvia Behling (Behling & Johnson Photography), pages 9, 13;
Russell A. Graves, pages 15, 17, 20; Charles Harrington (Cornell University), page 34;
David Lynch-Benjamin (Cornell University), page 31; Lynn M. Stone, pages 4, 6, 26,
29, 37, 42; SuperStock, pages 10, 18, 22, 25, 44; University of Wisconsin, page 39.